Handbook

MARINE CORPS INSTITUTE

AF807633

AFGHANISTAN: AN INTRODUCTION TO THE COUNTRY AND PEOPLE

MARINE BARRACKS
WASHINGTON, DC

Introduction

Purpose

With United States involvement in Afghanistan, knowledge of the country is imperative. This "handbook" explains a number of basic issues that should be in the knowledge "toolbox" of a Marine working with Afghanis or deployed to the country.

Three Forces

Three main forces have shaped present day Afghanistan: Tribalism, Islam, and geography. They have permanent and far-reaching effects on Afghan society and on anyone involved with that society.

Geography

Even though it is a landlocked country, Afghanistan has been on the crossroads of history. The country and people have been at the mercy of numerous empires and conquering states. Afghanistan was a prize worth possessing for its river basins and fertile valley and for its location on the east-west trade routes, especially the Silk Route to China. This geographic factor has affected and changed the peoples of Afghanistan.

Ethnic Interaction

The interaction of the various ethnic groups that invaded the region contributed to the patchwork of people that make up present-day Afghanistan. Each of these external groups that invaded affected the inhabitants of what is now Afghanistan in unique ways.

- For example, Genghis Khan destroyed most of the towns in the thirteenth century, forcing much of population into nomadism.

- A later Turkic dynasty, The Timurids, reconstituted the empire, but was forced out by pressure from newer groups on the west, east, and north of the country.

This constant shift and pressure of various external and internal groups has molded the Afghan attitude toward outsiders and indigenous groups of the country.

Continued on next page

Tribalism

Tribalism is another key factor that has affected the social organization of Afghan society. It affects kinship, loyalties, honor, and even revenge.

- The family as a unit can reach upwards of 50 people all united by the bond of a common "father."

- This tribalism, with emphasis on the male or father's family, means that sons often bring their brides to live in their father's home.

Afghanistan remains a patrilineal society, in which membership in both family and lineage, as well as property, is inherited only by kin on the father's side of the family. Kin on the father's side of the family also form close alliances.

International "Tribalism"

This tribalism is not just a local issue. Tribalism is what holds much of the Osama bin Laden's Al Qaida (Al Qaeda) terrorist network together.

- The bonding and the rules of social relations established in the village and nomadic setting play the same part in the international network.

- This tribal set of values that applies in the farming and nomadic settlements of Afghanistan and Pakistan has the same validity in Bosnia, Macedonia, and the United States.

Islam

Islam is the last influence that affects Afghanistan in a permanent way. It is a basis for the daily living and for understanding how Afghanis approach interpersonal relationships. It is a guide to the moral and ethical values that Afghanis hold as well as their spiritual mindset.

Continued on next page

Table of Contents

The table below lists the main topics and their page numbers.

Topic	See Page
Introduction	1
Quick Facts on Afghanistan	4
Geography and Climate	6
Economy and Livelihood	8
Languages and Ethnicity: Diversity	10
Languages of Afghanistan	11
Ethnic Groups: Pashtun	12
Ethnic Groups: Tajik (Tadjik)	14
Ethnic Groups: Hazara (Hezora)	15
Ethnic Groups: Uzbek	17
Recent Afghan History	18
Taliban's Rise to Power	20
Taliban: Who are they?	21
Taliban Relations with Neighboring Countries	22
Post-9/11 Afghanistan	23
New Government, Old Problems	24
Fighting Terrorism: Operation *Enduring Freedom*	26
Brief Military Chronology	30
Current Challenges	37
Rebuilding Afghanistan	45
Islam: Basic Ideas	47
Arabs and Islam	48
Muhammad (Mohammed)	50
Five Pillars of Islam	52
Al Qur'an (Koran) and Sunnah	54
Jihad (Djihad)	55
Islamic Schools (Sects)	56
Osama (Usama) bin Laden	58
Appendix A: History of Terrorist Activity	62
Appendix B: International Associations	63
Appendix C: Other References	65

Quick Facts on Afghanistan

Introduction	This brief resource guide covers a number of important subjects that give you a clearer understanding of the country. This page of quick facts is like a job aid. It highlights critical issues.
Land of Contrasts	Afghanistan consists of a major mountain range, the Hindu Kush, with its valleys, deserts and river system. http://fermi.jhuapl.edu/maps/afghanistan/ This territory includes many varied geographical and ecological zones. The different peoples of Afghanistan are as distinct and varied as the geography. Bringing them under one flag has made them into a "single" society.
Name	Transitional Islamic State of Afghanistan (TISA) as of December 2002 (formerly called the Islamic Emirate of Afghanistan by the Taliban in 1997) is a land-locked country in southwest Asia.
Size and Population	Area: 652,225 square kilometers or 251, 772 square miles. Estimated population: @ 27,000,000.
Population; Life Expectancy	0-14 years: 42.2%; 15-64 years: 55.01%; 65 years and over: 2.79%. Life expectancy about 46 years.
Languages and Ethnicity	Pushtu 35%, Afghan Persian (Dari) 50%, Turkic languages (primarily Uzbek and Turkmen) 11%, 30 minor languages 4%.
Religion	Sunni Muslim 84%, Shi'ite (Shi'a) Muslim 15%, other 1%.
Administrative Sectors	Country is divided into 30 provinces or velayets. The largest in area are Herat, Helmand, Qandahar (Kandahar), and Farah. The largest in terms of population are Kabul, Herat, Balkh, and Ghazni.
Population Centers	Principal towns include Kabul, Kandahar (Qandahar), Herat, Mazar-i-Sharif, Jalalabad, and Ghazni. http://www.afghana.com/GetLocal/Afghanistan/Provinces.htm

Continued on next page

Quick Facts on Afghanistan, Continued

Government/ Legal System	A transitional government, consisting of a grand assembly, president, and judiciary; no longer based on the Qur'an (Koran), the holy book of Islam.
Politics	Taliban (religious students) headed by Mullah Mohammad Omar was deposed in late 2001; a transitional government based on United National Islamic Front for the Salvation of Afghanistan or UNIFSA, often referred to as the Northern Alliance, and exiled leaders, forming a legislative assembly, a presidency, and a judiciary.
Economic Sectors	Agriculture: 53% of gross national product (GNP); industry: 28.5% of GNP; services: 18.5% of GNP (1990).
Additional Sites	http://www.odci.gov/cia/publications/factbook/geos/af.html -- CIA site that has handbooks of countries of the world. http://ebooks.whsmithonline.co.uk/htmldata/ency.asp?mainpage=HTTP://EBOOKS.WHSMITHONLINE.CO.UK/ENCYCLOPEDIA/01/C0000001.HTM -- short summary about Afghanistan from *The Hutchinson Family Encyclopedia*. http://www.nutshellnotes.com/afghanistan_text.htm – short dictionary-like definition information about the country.

Geography and Climate

Land-locked

Afghanistan is a landlocked country in southwest Asia. To the north, its neighbors are Turkmenistan, Uzbekistan, and Tajikistan. Iran is to the west; the People's Republic of China to the northeast, and Pakistan to the east and south. http://www.lib.utexas.edu/maps/cia01/afghanistan_sm01.jpg

---

Chief Topographic Feature

The Hindu Kush mountains are the main topographic feature that marks Afghanistan. These mountains are a barrier between the northern provinces and the rest of Afghanistan. They divide the country into three distinct geographic areas: the central highlands, the northern plains, and the southwestern plateau. You can find good maps of Afghanistan at http://www.lib.utexas.edu/maps/middle_east_and_asia/afghanistan_topo86.jpg

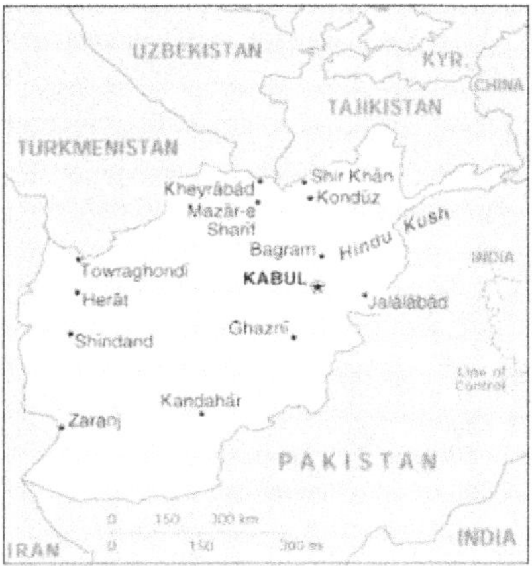

Source: http://www.globalspecops.com/oef.html

Continued on next page

Central Highlands

The central highlands are an area of about 160,000 square miles. This part of Afghanistan has deep, narrow valleys, as well as high mountains, which have proven to be historically important to the defense of the country. The Khyber Pass, one of the most famous routes to the Indian subcontinent, is located in the mountain ranges of the central highlands. http://www.maps.com/cgi-bin/search/hyperseek.cgi?search=CAT&Category=Asia%3AAfghanistanP&Qualifier=

The climate of the central highland is usually dry, with temperatures in the summer averaging around 80 degrees Fahrenheit; the winters are very cold.

Southern Plateau

This region of Afghanistan consists of high plateaus and sandy deserts. This desolate region covers about 50,000 square miles and is crossed by several large rivers. The average altitude of the southern plateau is about 3,000 feet. Kandahar (Qandahar), at an elevation of about 3,500 feet, enjoys a dry, yet mild climate.

Northern Plains

The northern plains are about 40,000 square miles of extremely fertile foothills and plains.

- The Amu River (formerly called the Oxus River) runs thought the foothills of this region.
- The average elevation is about 2,000 feet.

Agriculture is the main occupation in this region. There are large amounts of mineral deposits and natural gas in the northern plains.

Climate

The climate of Afghanistan varies dramatically between the mountain country and the lowlands. In the southwest during summer the temperature can reach 120 degrees Fahrenheit. In the Hindu Kush mountains during winter it can fall as low as –15 degrees Fahrenheit.

Map Site

http://www.lib.utexas.edu/maps/afghanistan.html -- This site has many different kinds of maps of Afghanistan.

Economy and Livelihood

Economic Areas

- Most people in Afghanistan are farmers or peasants who cultivate crops for personal use or for marketing and selling as produce. Herding is the next favored occupation by population size, followed by forestry and fishing.

- Manufacturing is the second largest area, followed by wholesale and retail commerce, then by transportation and mining. (With the current chaotic situation in the country, much of the economy has been disrupted.)

Agricultural Economy

Agriculture requires five major means of production: land, water, seed, labor, and animals (or tractors). Water is the most important resource in this respect. Each village has an official chosen by the elders or the village council. He regulates the division of water among landowners.

Family Farming

The family farm is the most common form of ownership and land cultivation. The family usually provides the labor and animals to cultivate its own land. Other agricultural labor arrangements include:

- Sharecropping, where the tenant receives a part of the produce.
- Commercial farming, where the owner pays cash to the farmers.

Because of the current chaotic conditions in Afghanistan, these agricultural systems are in decline.

Herding

Pastoral nomadism has been the traditional way that Afghanis have raised and herded their cattle, sheep, and goats. These nomads usually maintain their herds in groups, traditionally as clans or tribes. They might rely on the peasants or farmers for staples such as grain and implements they need for running a household.

Continued on next page

Economy and Livelihood, Continued

Mobility: An Advantage

Because of their mobility nomads were:

- The bearers of information and news.
- A traditional source of credit because their herds were a "countable" tangible source of cash. In the towns and villages where they took their stock to sell, nomads could buy goods on credit to sell to the isolated farmers or peasants living near their summer pastures.
- Often smugglers, both inside the country and across the borders.

Again, based on the chaotic war conditions of the country the usual nomadic herding patterns may have diminished or disappeared.

Manufacturing and Commerce

Besides food staples, such as margarine and wheat flour, woven fabrics, clothes and shoes are produced.

- Fertilizer, cement, and electric energy are the three main industrial products.
- Mining includes hard coal, gypsum, and natural gas.

The war in Afghanistan has caused industrial and commercial output to decline.

Languages and Ethnicity: Diversity

Introduction

As a landlocked nation, Afghanistan is at the crossroads for much migration and many settlements. The result of this exposure is a country with diverse ethnic and linguistics (language) groups.

Two Perspectives

- The next two sections give you an idea of the kinds of people living in Afghanistan, including the ethnic groups of the country. An ethnic group consists of people who have many cultural traits in common. These may include language, religion, morals, customs and ceremonies, economics and livelihood.

- Another way to look at diversity in a country is according to the languages the peoples of the country speak. Languages may cross ethnic boundaries, for example in Afghanistan, Dari is a common language to many of the different ethnic groups in the country. It crosses the ethnic/cultural boundaries of these different groups.

Languages of Afghanistan

Pushtu (Pashtu)	Pushtu is the official language of Afghanistan as, for example, Turkish is the official language of Turkey. Pushtu is also spoken in northern Pakistan. The language is written in a modified Arabic script.
Dari	Dari or Afghan Persian or Eastern Farsi is spoken by about 43% of the population. It is often referred to as the language of business and commerce.
Turkic Languages	About 11% of the population speaks a Turkic language, mainly Uzbek or Turkmen. Afghanis living close to the border of Turkmenistan and Uzbekistan probably speak Turkic languages.

Ethnic Groups: Pashtun

Introduction

Like the languages of the country, there are various ethnic groups in Afghanistan. The main groups are the Pashtuns (pronounced PAH-shtoons) (Pathans), Uzbeks, Hazaras (Hezoras), Tajiks (Tadjiks), and other minor groups.
http://www.lib.utexas.edu/maps/middle_east_and_asia/afghanistan_ethnoling_97.jpg

Definition

The Pashtuns live in southern and eastern Afghanistan in a population belt stretching from Kabul in the east to Kandahar (Qandahar) to Zaranj, which is on the border with Iran. They also live in western Pakistan.

- Pashtuns speak Pushtu (Pashtu) and are Sunni Muslims (see the section on **Islam** for an explanation of Sunnis). Many Pashtuns also speak Dari, the second language of Afghanistan.

- Pashtuns make up about 38% of the population of Afghanistan. Until the recent Taliban take-over, the Pashtuns controlled the ruling dynasty of the country for about 200 years.

Settlements

Pashtuns generally live in tribal villages. A number of villages may be near a town. A village is built primarily with regard to the location of water and to self-defense. The dominant lineages or extended families usually have the best and most strategic lands. Familial ties determine a group's location relative to the dominant tribe.

Nomads

Some Pashtuns are herders. They move from site to site following the pastures as the seasons change. Like the permanent villages mentioned above, the campsites of these nomadic Pashtuns are based on tribal lineages. Their tents are made from goat hair, supported by posts or arched poles.

Economy and Livelihood

Pashtuns work at grain farming and animal husbandry (see **Nomads** above). Wheat is the most important crop. Other important products are fruits, nuts, vegetables, and opium.

Industry usually includes trades such as shoemaking and carpentry. Some Pashtuns are small businessmen and traders.

Continued on next page

Kinship

The tribe is the unit of kinship. Descent is claimed from the father's side of the family.

- The clan is the principal division of a larger tribe. A clan is usually the offspring of one man, often including four or five generations, what is sometimes referred to as an extended family in Western terms. In this clan, a person marries and forms bonds.

- The smallest kinship unit is a household. The household or *kor* is the main economic and social unit; its members may form a village, single housing compound, or a nomadic group.

Local Authority

Pashtuns have a concept and practice of local authority based on tribalism.

- Honor (*nang*) is a tradition that pervades Pashtun society. It means there is little social class and no central authority.

- Taxes or rent (*qualang*) gives status and social class to the local people, many of whom are landowners. Qualang also means that there may be a "centralized" authority at the local level.

National Government

Generally, villages consider that the national government has responsibility for conscripting young men into the military and for taxation, both of which take resources away from the clan/tribe. Thus, villagers may have an attitude of neutrality toward, non-cooperation with, or defensiveness against the central government.

Important Customs

Pashtuns generally believe in and follow these customs:

- Purdah: Strict separation of the sexes in public
- Chadori: Women wearing the veil.
- Pakhtunwal: A mixture of blood-line revenge, hospitality to guests, defense of those who are in one's care, chastity of married women and caring for the weak

Ethnic Groups: Tajik (Tadjik)

Definition

Tajiks (pronounced TAH-jeeks) make up about 25% of the population of Afghanistan. They live

- In a large area north and west of Kabul (this area borders on the Republic of Tajikistan)
- Clustered around the city of Herat in western Afghanistan

Tajiks are also the principal inhabitants of the Republic of Tajikistan, a former region of the Soviet Union.
http://news.bbc.co.uk/hi/english/static/in_depth/world/2001/war_on_terror/key_maps/ethnic_groups.stm

Language and Culture

Tajiks are primarily Sunni Muslims. They speak the Tajik and Dari languages. Tajiks are believed to be descendants of the ancient Persians.

Economy and Livelihood

Since Tajiks in urban areas are well educated and wealthy, they make up the elite of Afghanistan. Their influence is in commerce.

- The plains-dwelling Tajiks live mainly in Herat, Parwan, and also around Kabul. They are town dwelling and are traders, skilled artisans, and farmers, all middle-class. Because they have settled in towns, they have replaced their tribalism with an urban orientation and a strong sense of community loyalty.

- Tajiks in rural areas make their living by farming and herding. They are a sedentary people. Landowners are usually village leaders.

Politics: Importance in Northern Alliance

Tajiks are the second largest group after the Pashtuns. They are also the Pashtuns' closest rivals for power and prestige, but with two exceptions, one in the 14th century and one for 9 months in 1929; the Tajiks have never ruled their region.

Tajiks formed an important part of the Northern Alliance, which opposed the Taliban. From a group of Tajik farmers, the chief leader of the Northern Alliance, Ahmad Shah Masoud, emerged. First he fought a guerilla war against the Soviet Communists and later against the Taliban. Taliban supporters assassinated him in late summer 2001.

Ethnic Groups: Hazara (Hezora)

Definition

Hazaras (pronounced HAH-zahrahs) make up about 19% of the Afghani population. They are Shi'ite Muslims, not Sunnis (for more details on Shi'ites, see the section entitled **Islam**).
http://news.bbc.co.uk/hi/english/static/in_depth/world/2001/war_on_terror/key_maps/ethnic_groups.stm

---

Groups/Settlements

The Hazaras fall into three groups: Settlements are located in an areas bounded

- On the north by the town of Bamian (Bamyan)
- On the west by Chaghcharan
- In the east by Kabul
- On the southeast by the town of Ghazni

The Hazaras themselves call this area Hazaristan, country of the Hazaras. Many Hazaras were forced off their land by the expansion of the majority Pashtuns and now live in Kabul and other towns.

Language

The Hazaras speak a language called Hazaragi. It is a mixture of Farsi (the language of Iran), Mongol, and Turkish.

Religion

- Because they are Shi'ites, the Hazaras have been subjected to oppression over the years. Most recently, the Taliban massacred about 300 Hazara men in January 2001. The reason was the Hazaras were suspected of collaborating with the United Front (Northern Alliance).

- Since they are Shi'ites, the Hazaras get support from Iran, which is predominantly a Shi'ite Muslim country.

http://news.bbc.co.uk/hi/english/world/south_asia/newsid_1500000/1500682.stm -- discusses Taliban persecution of Hazaras.

Continued on next page

Ethnic Groups: Hazara (Hezora), Continued

Origins

The origin of the Hazaras is interesting. Some researchers believe that Hazaras are the descendents of the 12th century Mongol warrior, Genghis Khan and his followers; others trace the Hazaras to the Buddhist civilization of 2000 years ago, the Bamiyan, whose center was in Afghanistan.

http://www.hazara.net/hazara/geography/Buddha/buddha.html -- an article about relationship of Hazaras to the Buddhist statues.

Economy and Livelihood

The primary occupations of the Hazaras are farming and trade. The main crops are wheat, barley, and tobacco. In some of the warmer areas, rice is cultivated.

Hazaras raise sheep, horses, and camels. Their horses are especially bred for hard work in the mountains. The wool from Hazara sheep is considered the best in Afghanistan. Camels are raised for their hides.

Social Organization

Like the Pashtuns, the Hazaras are organized by "tribes" or large clans. There are about 15 of these super-clans. The members belonging to a specific clan usually live in a certain geographic region. With so many clans, historically, there has been much inter-clan rivalry

Tribe/Clan: Examples

- **Daizinjat** clan lives in the area of Badsken, which borders on Herat and Koshak in the east. In Khila there are about 1,000 families. The Daizinjat tribe has about 30 clans. The most powerful is the Kaka clan.

- **Daikundi** consists of five tribes: Dolat Beg, Roshan Beg, Heider Beg, Gawash, and Barat. The greenest land of the Hazaras is in Daikundi, which is also called Hazarastan Kashmir. Before the Hazaras, the Tajiks lived in Daikundi.

Local and Central Governmental Authority

Because of their position as a religious and ethnic minority in Afghanistan, there has been constant tension between the local Hazara tribes/clans and the central authority of the country. The tension has boiled over a number of times into open revolution against the central authority, most notably in the early 1890s.

Ethnic Groups: Uzbek

Definition

The Uzbeks (pronounced OOZ-becks) are a Turkic-speaking ethnic group, which makes up about 6% to 8% of the population of Afghanistan. Most of them are Sunni Muslims. They are also the dominant population in the neighboring country of Uzbekistan.

http://news.bbc.co.uk/hi/english/static/in_depth/world/2001/war_on_terror/key_maps/ethnic_groups.stm

Location

Uzbeks live in a population belt that stretches along the northern border of Afghanistan from Maymaneh through Mazar-e-Sharif and Konduz. This Uzbek population belt parallels Afghanistan's northern border with Turkmenistan and Uzbekistan.

Social Structure

There are two types of Uzbeks: farmers and nomads. The farmers and city dwellers are mainly businessmen and tradesmen. The nomads raise sheep, goats, and cattle.

Settlements

The basic living unit is the village. Again, there is a nomadic village and a more permanent farming/commercial village. Both villages are based on kinship.

- The nomadic village is mobile, moving to pasturage as the seasons change.
- The farming village is a concentrated, permanent settlement: Houses are built close together. Most houses are walled and form a compound with a central courtyard.

Importance of Uzbeks in Afghanistan

Although a smaller minority in Afghanistan, the Uzbeks have played an important role in the country. One of the most important and most controversial groups who opposed the Taliban was the forces under the control of the Uzbek leader Abdur Rashid Dostum.

- Dostum had allied himself with the government in Kabul but later turned against it and joined with the *mujahidin* in 1992.

- He is currently "at odds" with the present-day government of Afghanistan. http://www.afghan-info.com/Politics/Afghan_Mujaheedin/Warlords.htm

Recent Afghan History

Introduction

This section emphasizes recent history of Afghanistan to show the events that led up to the Taliban taking power in the country.

http://nths.newtrier.k12.il.us/academics/social/conflict/4s%20Web%20Pages/Period9/Afghanistan/afghanhome.htm -- a more detailed and lengthier history of the country.

Communist Take-over

- In 1978, during the Saur Revolution, the Afghan communist party under Muhammad Taraki overthrew the republican government. This communist party, called the People's Democratic Party of Afghanistan, started land redistribution, secularized the country and began to re-educate the people.

- Factionalism caused the party to break down. This allowed the Soviets to invade in 1979. Eighty thousand Soviet troops invaded the country. The Soviets replaced Taraki with their puppet Babrak Kamal.
   http://www.afghan-info.com/History/Polevent.htm#saour

At this point groups of Islamic tribes/clansmen formed a loose alliance that collectively became known as the *mujahidin*. They were anti-Soviet and anti-Communist.

Soviet Domination/ Rise of *Mujahidin*

During the 10-year occupation of Afghanistan, about five million Afghanis fled the country; close to 1.5 million were killed.

- During this period, the *mujahidin* gained prominence as the group that fiercely opposed the Soviets. They were supported by the West and by Muslim countries in their struggle against the Soviets and the puppet government that the Soviets had imposed on the country.

- The Soviets tried a policy of reconciliation with the opposition *mujahidin* in 1987, but the seven-party *mujahidin* alliance refused to observe a cease-fire and to participate in negotiations with the Soviet puppet regime.

Continued on next page

Soviet Domination/ New Constitution

A new constitution was adopted in July of 1987, allowing the formation of other political parties beside the communists.

- The communist party of Afghanistan said it would be willing to share power with opposition groups in a coalition government.
- The name of the country was changed back to the Republic of Afghanistan from the Peoples Democratic Republic of Afghanistan.

Soviet Withdrawal

The Soviets withdrew their troops in 1989. The last president, Najibullah, remained in power. Civil war continued as the *mujahidin* were strongly against the government of President Najibullah.

***Mujahidin* Rule: New Government**

A number of attempts were made to integrate the *mujahidin* in the central government through advisory councils of tribal leaders.

- Eventually Rabbani was elected president of the country.
- In 1992, the Peshawar Agreement was ratified: A purely Islamic interim government was installed in place of Najibullah's.

New Rules for Afghan Society

- Leadership in this new government rotated among the seven Islamic *mujahidin* groups.
- This coalition government banned alcohol and gambling, required women to wear the veil, and separated education into male and female institutions.

Factionalism

Infighting among the various factions of the *mujahidin* continued, even though each faction had two ministerial posts in the government.

- Eventually the alliance broke down into civil war. The most serious conflict was between Premier Hekmatyr, a Pashtun, and President Rabbani, a Tajik, together with his defense minister, Ahmad Shah Massoud.
- The result of this armed conflict was the almost total destruction of Kabul and more important, Pakistan's support for a group of Pashtun religious students, the Taliban.

Taliban's Rise to Power

Early Advances

In September of 1994, the Taliban (meaning students in Persian), an "army" of former Muslim seminarians and religious students, entered the civil war. By 1995, the Taliban had claimed victory in the town of Herat and had advanced on Kabul, the capital.

During 1995, the Taliban made extensive gains in the country, particularly in the south. In many instances, they were welcomed by the local Afghanis who were tired with and disgusted by the chaos, corruption, and instability that mujahidin factionalism had caused.

Occupation of the Capital

- By autumn of 1996, the Taliban had driven President Rabbani out of Kabul, captured the capital, and executed Najibullah. In May 1997, after a brief alliance between opposition forces and the Taliban, violent civil war again broke out.
- Fighting then continued between the United National Islamic Front (Northern Alliance), consisting of the *mujahidin* factions, and the Taliban.

Afghanistan divided: Taliban

The Taliban held about 85-90% of the area of Afghanistan. http://europe.cnn.com/SPECIALS/2001/trade.center/afghan.civil.html , http://www1.cnn.com/WORLD/9610/04/afghan/index.html -- good discussion of the problems the Taliban have in uniting the country.

United Front/Northern Alliance

- The Northern Alliance held the remaining 10% to 15% of the country. As of late November 2001, the alliance had extended its control over the country, capturing Kabul.
- The alliance was strongly against the Taliban's extreme fundamentalist methods and religious philosophy and was very much opposed to Pashtun ethnic dominance. http://www.washingtonpost.com/wp-srv/nation/graphics/attack/zone_3.html

http://www.afghan-web.com/politics/parties.html – lists anti-Taliban parties/groups and their leaders.

http://www.afghan-web.com/ciriello/alliance/ -- pictures of alliance forces.

Taliban: Who Are They?

Education and Training

The Taliban were originally young Afghanis many of whom went to Pakistan to study in madrassas (religious schools). These schools:

- Focus on instruction in the Qur'an; however, a number of them teach military training and militant Islam to their students.
- Were the source of the Taliban's interpretation of Islam and their approach to restructuring Afghanistan socially and politically. (Saudi and Pakistani intelligence services also trained these Afghan religious students.)

With their "purist" Islamic view, the Taliban were considered a group that promised a clean, uncorrupt society for Afghanis and were opposed to the factionalism and corruption of the *mujahidin*, who had plunged Afghanistan into civil war after the withdrawal of the Soviets.

Outlook

The Taliban are very provincial and backward in their outlook: The external world, much of which they considered unclean and corrupt, begins at the border of Afghanistan.

The Taliban have a close relationship to Osama bin Laden, whom they feel follows the same fundamentalist precepts they believe in. Their relationship to bin Laden was further cemented by the marriage of their leader, Mullah Omar, to one of bin Laden's daughters.

Taliban Oppression

Besides the requirement for women to be fully veiled and accompanied by a man in public, the Taliban had

- Stopped all education for girls and women
- Forbidden women to work
- Forbidden women to wear white socks
- Required men to grow beards
- Stopped the Afghan national sport of Buzkash
- Changed the name of the country to the Emirate of Afghanistan
- Forced conscription upon the male population of Afghanistan to increase their fundamentalist army

The law of the country was the Taliban interpretation of the Sharia, Islamic religious law, with strict penalties, such as death and stoning, for infringements of the various oppressive laws imposed on the Afghan population.

Taliban Relations with Neighboring Countries

Iran

Tensions between Taliban-controlled Afghanistan and Iran were always high. The Iranians were angered by the murder of their diplomats and a journalist by the Taliban near Mazar-i-Sharif. In response to these killings, in September 1998, Iran deployed 70,000 troops to conduct military exercises near the Afghan border, but no fighting occurred between the two nations.

- The Iranians also feel a "kinship" to the Shiite Hazaras of Afghanistan, whom the Taliban have persecuted and massacred.

- Finally, the extreme primitiveness and barbarity of Taliban fundamentalism alarmed Iran. The Iranians were probably glad to see the Taliban removed from Afghanistan as the controlling government authority.

Pakistan

The Taliban were supported among some groups of the Pakistani population and by some Afghan refugees living in Pakistan, particularly those close to the Afghan-Pakistani border and those trained in the militant madrassas. The kind of fundamentalism preached by the Taliban is also prevalent in some sectors of Pakistani society.

Although the government of Pakistan supported the U.S. effort to root out bin Laden and his network of terrorists based in Afghanistan, Pakistan still recognized the Taliban as the legitimate government of Afghanistan long after the rest of the world denied their legitimacy.

Other Countries

Tajikistan, Uzbekistan, and Turkmenistan were all apprehensive about the Taliban, particularly their type of Islamic fundamentalism. Many of these former regions of the Soviet Union, now independent countries, are struggling to establish democratic processes for their populations. They viewed the spread of Taliban Islamic fundamentalism as a challenge to their own move toward popular sovereignty.

On the other hand, since there are minorities from these countries, particularly Uzbeks and Tajiks, living in Afghanistan, the governments of these neighboring countries had to be careful about the amount and intensity of their anti-Taliban support.

Post-9/11 Afghanistan

9/11: Activating the US War on Terrorism

September 11, 2001 is a date that saw the United States and eventually other coalition countries begin a concerted effort to actively combat international terrorism, primarily represented by Osama (Usama) bin Laden. He was eventually placed on the FBI's "Ten Most Wanted List" with a $5 million reward for information leading to his arrest.

After 9/11, the U.S. government requested that the Taliban government of Afghanistan turn over bin Laden, the international terrorist who had masterminded the terrorist attacks on the World Trade Center (WTC) and the Pentagon.

- When the Taliban refused to extradite bin Laden, a US-led coalition was formed and after six weeks of aerial bombardments and military action on the ground the Taliban was ousted from power on November 17, 2001.

- Many Taliban dispersed to mountain retreats in the country or fled across the border into neighboring Pakistan. Osama bin Laden is still at large and his whereabouts are not known. The coalition hunt for these remaining terrorists and for bin Laden continues. (A brief discussion of military action in Afghanistan, including Marine Corps participation, is in **Fighting Terrorism: Operation Enduring Freedom** and **Brief Military Chronology**, both later in this handbook.)

Planning for a New Afghan Government

- In late 2001, important leaders of the anti-Taliban groups in the country and leaders from the Afghan diaspora [Afghanis living in exile abroad] met in Bonn, Germany.

- The groups signed onto the Bonn Agreement, a plan to create a new government in Afghanistan.

- A judicial commission was also established to rebuild the justice system according to Islamic principles, international standards, the rule of law and Afghan legal traditions.

New Government, Old Problems

Afghan Interim Authority

The beginning of the new Afghan government was the Afghan Interim Authority (AIA), which was established, with Hamid Karzai as its chairman on December 22, 2001.

International Pledges

In January 2002, to support the new Afghanistan government, sixty countries and international financial organizations present at the Tokyo Donors Conference pledged donations totaling $4.5 billion through 2006 for the reconstruction of the country.

TISA

The Afghan Interim Authority (AIA) held a nationwide grand assembly, called the Loya Jirga, in June 2002. Karzai was elected president of the Transitional Islamic State of Afghanistan (TISA) for a two-year term by the Loya Jirga. The TISA has:

- 18 months to prepare for and hold a Loya Jirga [grand assembly] to adopt a constitution for Afghanistan.
- 24 months in which to hold national elections.

In December 2002, the TISA marked the one-year anniversary of the downfall of the Taliban regime.

http://www.nutshellnotes.com/afghanistan_text.htm has a good update on Afghanistan, which is succinct and accurate.

Continued on next page

Continued Factionalism

Factionalism is still a critical issue in Afghanistan even though there is one transitional government. The split between the ethnically dominant Pashtuns, some of whom were Taliban, and the "minority" tribes of northern Afghanistan, such as the Tajiks and the Uzbeks is a hindrance to real unity.

- Besides this "tribal" rivalry, foreign involvement by groups vying for oil reserves complicates the problem of unifying the country.
- Other countries still fear the spread of residual fundamentalism left from the Taliban and object to possible civil rights abuses in present-day Afghanistan.
- The opium trade in Afghanistan also contributes to the factional problem.

Political Parties

Afghan political parties are not stable either. Karzai's government is a coalition of leaders from all political and ethnic groups. Various leaders in the country are planning to create new parties. Some of these political factions in the current government have formed new groups and parties, intending to participate in the 2004 elections.

Fighting Terrorism: Operation *Enduring Freedom*

Definition

Operation *Enduring Freedom* is the name given to US combat operations against terrorism in Afghanistan. *Enduring Freedom* began on October 7, 2001.

Purposes

The primary purpose for Operation *Enduring Freedom* is to seek out Osama bin Laden, who had been given sanctuary by the Taliban.

The Taliban who had controlled Afghanistan had allowed *Al Qaida* (Al Qaeda), the key terrorist group lead by Osama bin Laden, to establish training and base camps in Afghanistan. Thus another important reason for Operation *Enduring Freedom* is to root out and destroy these training and base camps and bring in-country members of Al Qaida (Al Qaeda) to justice.

A third purpose is to seek and destroy Taliban and anti-government forces operating in Afghanistan. As mentioned earlier, many Taliban members dispersed to mountain retreats when the coalition invaded Afghanistan. These residual fundamentalist militants are the focus of coalition search and destroy missions.

Another important reason is to ensure the stability that is necessary to reconstruct Afghanistan politically, economically, and socially. Without a stable environment, any new Afghan government will have a very difficult time uniting the country under one central government authority.

Continued on next page

Key Elements

Sea based: three strike aircraft carriers, each with various fighter squadrons aboard; the USS *Kitty Hawk*, which transported Army Special Operations helicopters; two amphibious groups built around the 15th and 26th MEUs, both of which later became the main US ground force in southern Afghanistan.

Diego Garcia base, which essentially housed B-1 and B-52 bombers.

Land basing in neighboring countries: Uzbekistan allowed US helicopters to operate out of Khanad; a few based in Tajikistan; Pakistan offered bases covertly out of Pasni and Jacobabad. These bases enabled the special forces assault on 17 October and the Marines' forward operation base (FOB) Rhino.

Precision Air Attack: Joint direct attack munition (JDAM) which flew to preset coordinates based on satellite information (GPS/INS). Wind-corrected munitions (bomblet) dispenser (WCMD) first used in Afghanistan; laser-guided bombs. Electronic sensor surveillance used unmanned air vehicles.

A big plus was the fact that heavy bombers could engage multiple targets with JDAMs.

NATO Involvement

On September 12, 2001, the North Atlantic Treaty Organization (NATO) met in emergency session and, for the first time in its history, invoked Article 5 of the Washington Treaty. This article states that an attack against one NATO country is an attack against all member countries. As a result:

- All 19 NATO Allies and the 9 countries applying to join NATO have provided blanket overflight rights, access to ports and bases, refueling assistance, and increased law-enforcement cooperation.
- NATO Airborne Warning and Control System (AWACS) crews logged over 3,000 hours patrolling and protecting American skies while U.S. planes flew missions in Afghanistan.
- Sixteen NATO allies now support Operation *Enduring Freedom* in Afghanistan and the global campaign against terrorism.
- Fourteen allies have deployed forces in the region.
- Nine allies are participating in combat operations in Afghanistan.

Continued on next page

NATO Involvement, continued

- Allies and other partner countries have deployed nearly 4,000 troops to Afghanistan and have also provided 95% of the international security assistance force (ISAF), led first by the United Kingdom and now by Turkey.
- As of August 2003, NATO took control of all security in Kabul. This is NATO's first operational commitment outside Europe.

See more details at: http://www.globalspecops.com/nato.html

Results

Twenty-seven days after the 9/11 terrorists attacks on the World Trade Center and the Pentagon, the US joint team invaded Afghanistan and two months later freed Afghanistan from Taliban domination, together with US allies.

This was accomplished without:
- Any US bases in-country
- A major plan for removing the Taliban from power.

http://www.lejeune.usmc.mil/2dmardiv/36/ has detailed reports and lessons-learned from Marne Corps experience in Afghanistan. Click on the link above; click on operations, then on professional forum. Then click on any of the reports individually.

Continued on next page

Key Accomplishments

- The deepest amphibious operation in Marine Corps history – over 400 miles. Task Force 58 conducted operations in Afghanistan far from the sea. The operations showed that Marines could operate expeditionary forces at extended distances from a sea base.

- Marines and soldiers fought battles at extremely high elevations, over 10,000 feet above sea level.

- *Enduring Freedom* was the first time C-17's were employed in a medium threat environment to air drop supplies.

- For the first time, headquarters and the aircrew viewed a picture taken from an unmanned vehicle simultaneously over the target.

- Operation *Enduring Freedom* included the US Air Force's longest combat sortie on record – 44 hours. Over all, naval aircraft flew about 75% of all sorties and dropped about 33% of the bombs. The Air Force flew about 25% of the combat sorties and dropped about 75% of the bomb tonnage, including the "Daisy Cutter."

Brief Military Chronology

Introduction

This section is an abbreviated discussion of military action during Operation *Enduring Freedom*. The topics include the following:

Topic	See Page
Early Days	30
War in the North	31
War in the South – FOB Rhino	32
Capturing Kandahar	34
War in the South: Khowst-Gardez Mission	34
Tora Bora and Operation *Anaconda*	35

Early Days

On October 7-8, 2001, The United States and Great Britain began strikes against Taliban air defenses, fuel sites, al Qaida (Al Qaeda) terrorist training camps and headquarters and military installations.

- The main thrust of air assaults was the Taliban regime in Afghanistan: Concurrently crippling Taliban movement, the allies supported the Northern Alliance

- Such precision bombing only stiffened the Taliban who did not hand over Osama bin Laden.

These targeted actions were designed to disrupt the use of Afghanistan as a terrorist base of operations and to attack the military capability of the Taliban regime. At the same time the U.S. and its allies began to drop food, medicine, and supplies to the population of Afghanistan.

Continued on next page

Brief Military Chronology, Continued

War in the North

Two issues moved the war effort forward in northern Afghanistan: increased bombing by the allies and the commitment of US Special Operations Forces (SOF). This same kind of commitment was to be important in the success of the Marines later in their search for residual Taliban and Al Qaida forces.

- On November 9, Mazar-i-Sharif fell to the Northern Alliance, effectively opening a land route directly to Uzbekistan and blocking any northward movement of the Taliban from Kabul to Uzbekistan. Capturing this city also offered coalition forces a major airfield in-country.

- The momentum of the Northern Alliance supported by the allies rolled on until the Taliban abandoned Kabul on 12 November.

- The last Taliban stronghold, Konduz, surrendered on 26 November.

War in the South – FOB Rhino

Background

The Taliban had a stronger hold on southern Afghanistan. It was more difficult in the south to muster the local population to oppose them than in the north. The Taliban also had better security forces there.

MC Mission

The plan called for TF58 to seize and hold Rhino, while conducting raids from it. As General James N. Mattis, commander of TF58, expressed it:

> *"...to create chaos, denying the enemy their sense of security."*

This coupled with guidance from Vice Admiral Moore, Commander NAVCENT/ MARCENT: The Marine raids would defeat the Taliban and Al Qaida forces quickly and decisively became the Marines' mission.

Elements

US Army Special Forces: Generally, these forces could initially seize an airfield or installation, but did not have the manpower to hold it. This is where Marine Corps forces came into play.

Marine Corps: Task Force (TF) 58 consisting of the
- 15th Marine Expeditionary Unit (SOC) [MEU (SOC)]
- 26th MEU (SOC).

Location

An airstrip that the Special Forces had raided, called FOB Rhino.

Continued on next page

Establishing the FOB

Marines seized FOB Rhino on November 25, 2001.

- They linked up with Navy SEALs who had earlier arrived there for target surveillance and reconnaissance.

- Once the airfield was declared KC-130 capable, additional Marines were landed. The MEU's attack helicopters provided close air support.

TF 58: Interdiction of Rt. 1

This operation began on December 4, 2001, and included an approximately 85-mile, 19-hour motorized movement across the desert. The raid resulted in the:

- Capture of a Taliban convoy.
- Detection of a P-3.
- Killing of 120 enemy personnel.

Capturing Kandahar

Background

Kandahar was a Taliban stronghold, the center of their power. The Taliban refused to surrender the city to Afghan Southern Alliance forces, and Al Qaida began to slaughter local Afghanis who were surrendering. Finally, on December 6, 2001, the city fell.

MC Action at Kandahar Airport

TF58 linked up with the Southern Alliance and U.S. Special Forces on December 13, 2001.

- By December 14, TF58 had established a second FOB at the Kandahar airport.
- Marines repaired the airstrip so that by December 18 the airport was considered safe for C-17s.

Further MC Operations

Looking for intelligence materials:

- In villages west of FOB Rhino
- Near Kandahar

War in the South: Khowst-Gardez Mission

Background

Although Kandahar had fallen to the alliance and coalition allies, strong pockets of resistance still remained. The local alliance leader had merely raised the alliance flag; his troops had not actually taken on the Taliban.

TF58's Mission

TF58 was tasked to seek out and destroy these pockets. Special Forces would identify the targets, and Marines would "hammer" them. Because the distance from Kandahar to the battle space was about 250 miles, a local airfield, about an hour from the battlefield, was used for close air support.

Disestablishing TF58

During January 2002, TF58 was relieved in place by troops of the Army's 101st Airborne Division; however, when the Taliban attacked the Kabul airport, the Marines beat them back. TF58 was disestablished at the end of February 2002.

Tora Bora and Operation *Anaconda*

Al Qaida (Al Qaeda): Two Options for Escape

Al Qaida, including Osama bin Laden, and surviving Taliban leaders and forces had had two possibilities for escape from invading coalition forces:

- They could leave Afghanistan by escaping to Pakistan or through Pakistan to countries that harbored groups sympathetic to the Al Qaida cause, such as Somalia, Indonesia, or the Philippine Islands.

- The second possibility was to go into hiding in the mountain caves and retreats in the Tora Bora Mountains in eastern Afghanistan, near the Pakistani border.

Tora Bora

The assault on Al Qaida (Al Qaeda) in Tora Bora was conducted by Afghan troops strongly supported by U.S. firepower. Few U. S. troops were directly involved in the assaults. Afghan troops took the caves, but many of the key Taliban and Al Qaida figures there escaped.

About 300 bodies of bin Laden fighters were counted; other bodies were probably crushed beneath collapsed caves and bunkers. 150 more enemy personnel were caught trying to escape across the border into Pakistan.

Tora Bora taught Al Qaida forces that U. S. detection systems could intercept messages from radios and cell phones. Al Qaida would attempt to profit from this lesson-learned in Operation *Anaconda*.

Operation *Anaconda*

Operation *Anaconda* (in Gardez and Shah-e-Kot Valley) was mainly a U. S. forces operation. British, Canadians, and Australians also took part in the action. There were about 1,000 Afghan troops, 1,000 U. S. and some additional coalition members.

The purpose of the operation was to destroy what was considered the last pocket of Al Qaida troops, thus preventing them from launching an offense the following spring.

Continued on next page

Tactics

Limited attacks on Al Qaida (Al Qaeda) troops to force them to reveal their positions so that air strikes could be called in.

- Apparently, Al Qaida troops attacked whenever they saw U.S. troops, regardless of the conditions of the battlefield, e.g., they would attack uphill and into fire.
- Al Qaida also had little understanding of the sensors and sophisticated weapons they faced.

Cave Raiding: Results

Although there were some problems, in general, this "cave raiding" was worthwhile:

- Documents and computers, with clues to future Al Qaida (Al Qaeda) operations as well as lists of Al Qaida personnel and financial accounts of the group were recovered.
- Information about the Al Qaida's interest in weapons of mass destruction was also forthcoming.

Protecting the interim Afghan Government

In December 2001, The United Nations assembled the International Security Force (ISAF) to stabilized Kabul, the capital of Afghanistan. The bulk of the ISAF came from Muslim troops from Turkey and Jordan, with small contingents from the Canada, Western Europe and Australia and New Zealand.

Note: Much of this brief military chronology came from chapter 9 "Striking Back" of Norman Friedman's Terrorism, *Afghanistan, and America's NEW WAY OF WAR.*

Current Challenges

Introduction

Rather than try to explain the current challenges of deploying to Afghanistan, a number of "case studies" are detailed below. You should be able to get some idea of the "working" environment in country that Marines have already encountered and that you may have to face. The topics in **Current Challenges** are shown in the table below:

Topic	See Page
Local Problem: Low Intensity Civil War	37
Warlord Case Study	38
MC Security Squad Experiences: Kandahar	39
MC Security Squad Experiences: Raids near Khowst	41
Joint Operations	44

Local Problem: Low Intensity Civil War

"*Karim Khan stands disconsolately outside the local government headquarters in the remote village of Tuksar* [in Afghanistan]. *He used to run the neighbouring village, but was bundled out by a rival militia one night recently, leaving his wife and family behind as virtual prisoners.*"

- The incident is not isolated. It is not so uncommon throughout northern Afghanistan, where central government control is not as strong as it is near the Afghan capital Kabul.

- This has led a number of experts to consider some aspects of the combat situation in Afghanistan today as low-level civil war in which local militias use the autumn, the country's traditional fighting season, to change the map of power.

Continued on next page

Feuding Warlords

Government curbs northern warlords

Feuding commanders' armies to merge as Kabul shows resolve.
By Ahmad Nahim Qadiri in Mazar-e-Sharif (ARR No. 79, 29-Oct-03)

"Afghanistan's transitional government has taken decisive action to improve security in the north of the country, dismissing key officials in Balkh province and ordering two feuding military forces t merge. In a major shake-up announced by interior minister Ali Ahmad Jalali on October 26, the governor and deputy governor of Balkh province were replaced and the chief of police in the administrative centre Mazar-e-Sharif was dismissed.

"Jalali also announced that the two army corps run by rival commanders General Abdul Rashid Dostum and General Atta Mohammad are to be merged into a single force and integrated into the new Afghan national army. Disarmament of heavy and light weapons is scheduled to begin in a few weeks in Mazar-e-Sharif, supervised by the national defence ministry, the United Nations Mission in Afghanistan, UNAMA, and the British-led provincial reconstruction team, PRT.

"Analysts in Mazar-e-Sharif say the changes are designed to strengthen the hand of President Hamed Karzai's government in the north of Afghanistan, ahead of crucial decisions on a new constitution and elections planned for next year. They say the decisions are important for symbolic . well as practical reasons. ... The changes come amid heightening tension in Mazar-e-Sharif, where there were fears that a ceasefire agreement between the forces of Dostum and Atta Mohammad could collapse.

"Dostum, who leads the Junbesh-e-Milli movement and has been a key player in the north for more than a decade, has been battling against Atta Mohammad for control of Mazar-e-Sharif and surrounding territory since the fall of the Taleban regime nearly two years ago.

"The two men control large forces -- designated the Eighth and Seventh Corps, respectively – which formally come under Defence Minister Mohammad Qasim Fahim, who like Atta Mohammad is part of the powerful Jamiat-e-Islami faction. The turf war between the two groups has frustrated attempts by the transitional government to extend its authority to the northern provinces."

As you can see, this article points up a number of issues that in-country deployed forces may have to deal with:

- Indigenous forces that are not united
- Tension between the central government and locally based leaders
- Local vice the national view on demobilization
- Turf-war claimants

From:
http://www.reliefweb.int/w/rwb.nsf/480fa8736b88bbc3c12564f6004c8ad5/2473c6b865d78c8849256dcf00120fa8?OpenDocument

Continued on next page

Introduction

As a member of 3d Bn, 6th Marines, L Co., 1st Plt., 26 MEU (SOC), these are the experiences of an infantry platoon sergeant who served in Afghanistan:

My platoon was the security element for the company. We arrived in Afghanistan on 19 Dec 2001 and stayed until 5 Feb 2002.

FOB Rhino: Insertion

- We flew from Pasni, Pakistan, into FOB Rhino on 19 Dec 2001. We went from 90-degree heat to 40-degree chill in 3 hours. What an adjustment! After the first couple nights we became accustomed to the drastic temperature change.

- The biggest challenge was the feeling that we were isolated. Since 1/1 was securing the outer perimeter we only had to man the interior posts. After 2 days we were flown by C-130 to Kandahar.

Kandahar

Purpose

Our purpose at Kandahar was to augment security of the airfield:

- We didn't have much interaction with the civilians.
- The civilians we did meet seemed very pleasant and not hostile in any way.

Our translator told us the people were very glad to have security and stability in their lives for once.

Climate

The temperature in January ranged from 50° F. in the day to 15 °F. at night. You can expect a 20-30 degree temperature change when the sun sets.

It was dry except for two rainy days. When it rains, the sand and hard packed dirt become ankle-deep mud. The mud is slippery and it stinks!

Continued on next page

Kandahar (Continued)

Keeping Warm

Marines will find ways to stay warm no matter what. Here are a few ways we kept warm.

- While on watch Marines would lay their poncho liners over their laps.
- Hot wets were prepared before nightfall and served out to the Marines.

These little things plus the gear we brought with us made our nights passable.

Keeping Clean

Since we didn't have showers, and water was at a premium, we had to take canteen cup baths, better known as birdie baths.

- We would heat up a canteen cup full of water and wash ourselves head to toe.

- With that same canteen cup of water we would then wash our socks and tie them to our packs to dry in the daylight.

After a few weeks we received wash buckets to wash our clothing. Teaching Marines how to wash clothes by hand and with a washboard was unique!

Making Life Bearable

We had packed enough tobacco products to last 30 days. Since the army took nearly a month longer to get there, we made arrangements with some air force personnel to bring some tobacco in to us once in a while.

Mail didn't arrive in the country until the week before we departed. Priorities were water, food, equipment, ammo, fuel, and then mail. However, the MEU did set up three to four phones for us to use. We would have to wait for hours to make a call, but it was worth it.

Coffee was made by pulling a clean sock tip around the top of a canteen cup; put the coffee grounds on the sock, then pour hot water over the grounds. Coffee made a huge difference!

Continued on next page

Kandahar (Continued)

Dangers

While in Kandahar there were a few things we, Marines, always kept in the forefront of our minds: mines, mines, and more mines.

- Not only did we have to watch for Soviet mines, we had to be careful of our own cluster bomblets. Wherever we saw some in the area, we had our Marines take a look at them so they would know what they looked like.
- Unexploded ordinance was everywhere. EOD was busy with the bigger stuff. All our Marines could do was mark near where the ordinance were and avoid the area.

Important!

Some precautions that paid off:

- Gear inspections and gear accountability (No one lost any gear!)
- Uniformity within the company
- Entrenching was continuous. We made bunkers not just fighting positions.
- Weapons training (not one ND while we were there!)
- Training the NCOs (Never did I catch a Marine breaking light discipline or falling asleep on watch) The NCOs ensured their Marines were alert and prepared!

Raids near Khowst

Purpose

Our purpose near Khowst was to conduct a raid on cave complexes located near there. The raid was supposed to last under 10 hours from insertion to extraction; however, it lasted 10 days.

- In those 10 days we gathered much of our intelligence information.
- We also proved that Marine units could and did survive, fight, and prevail without resupply.

Climate

Khowst is located in northeast Afghanistan, near the border with Pakistan. The climate in the mountains at night was extremely cold, in the teens to single digits. During the day the temperature was 40° F-50° F. There were trees and scrub brush in the area. It didn't rain during the time we were up there.

Continued on next page

Raids near Khowst (Continued)

Terrain

The terrain is mountainous. Physical preparation and education is needed for operating in a mountainous environment.

- Vehicle movement is restricted to the *Wadis* and roads.
- There are many old bunkers, fighting positions and caves throughout the mountains.
- There were trees and scrub brush in the area.

Enemy Tendencies

The enemy made bivouac sites opposite their caves. The cave locations were mostly in the *Wadis* where they were extremely difficult to see or bomb.

- We found equipment stashed in burial sites, buried outside villages, and just left plainly in the open.
- The enemy signaled at night from mountaintop to mountaintop with white light.
- There was also evidence that they had cell phone/satellite phone capability.

As soon as we landed, we found evidence that the enemy had hastily withdrawn into Pakistan. After a few days we had a helicopter arrive to take detainees to Kandahar. Some started making their way back over the mountains.

Interacting with the Local Population

Some of the people we took as detainees were absolutely scared of us. Based on our physical size and appearance, they would become almost hysterical, but with a little show of kindness on our part, they willingly told us many tidbits of information: All we had to do was give them a Koran and let them know when it was prayer time. All of a sudden we discovered that one of them could speak English.

- These people knew which group controlled what area; they would claim allegiance to both so that they could have unfettered access to both areas.

- The people seemed to side with whoever was more powerful.

- They have little national pride; they did have pride in their tribes!

In a vehicle we captured going down the *Wadi* we found a Taliban flag and a Northern Alliance flag. In the same vehicle we also found a GPS.

Continued on next page

Raids near Khowst (Continued)

Living off the Land

Because we had no re-supply, we had to live off the land (a 10 hour-raid means you carry very little gear).

- We seized an abandoned terror training facility and found livestock and grain.
- We soon began making corn bread, pop corn, and other types of food. After a couple days we had chicken, lamb, and steak.
- We found water in cisterns and in natural springs in the *Wadi*. We purified the water in 5-gallon jugs.

Small Unit Operations

In the mountains we had great success, operating at squad size and less. There were always 2 patrols out day and night. They would range 3-4 km and stay at altitude. This helped us gain vital information of enemy movements and discover weapons caches.

We constantly used binoculars and NVGs with 3x Mag for observation. This timely warning and information meant we could capture many people who may have otherwise become combatants.

http://www.lejeune.usmc.mil/2dmardiv/36/ has detailed reports and lessons-learned from Marne Corps experience in Afghanistan. Click on the link above; click on operations, then on professional forum. Then click on any of the reports individually.

Continued on next page

Broad Tactical Environment

By Jim Garamone

American Forces Press Service
WASHINGTON, June 4, 2003 - U.S. and Italian forces launched an offensive June 2 against anti-coalition forces in the Shahi Kowt section of the Paktia province in Afghanistan, Combined Joint Task Force 180 officials said today.

A brigade-sized unit launched Operation Dragon Fury to prevent the re-emergence of terrorism, to deny sanctuary to terrorists and to protect nongovernmental organizations and coalition forces from enemy attack.

Army Col. Rod Davis, spokesman for the command, said the ground assault section set up blocking positions and air assault troops drove anti-coalition forces into the trap. A total of 21 men were taken under control by coalition forces.

The U.S. forces came out of Kandahar, the Italian forces from Khowst. Paktia province is located in eastern Afghanistan near the border with Pakistan. The province was the scene of tough fighting in 2001 and was raided again last year.

Roughly 20 helicopters launched the air assault. Close-air support was standing by in case the coalition troops needed the assistance, Davis said. He said the Italian forces searched more than 300 vehicles and inspected more than 800 people.

There were no coalition casualties.

American forces came from the 82nd Airborne units in the area. Special operations personnel also took part.

From: http://www.globalspecops.com/oef2a.html This article shows the "larger environment" in which Marines like those in the earlier case study might be involved in.

Rebuilding Afghanistan

Armed Forces

The United States is helping the Afghan government to build a national army. In 2002, US soldiers helped train 1,600 Afghan soldiers for this new army.

http://www.globalspecops.com/rebuilding.html is good site that summarizes US efforts to rebuild Afghanistan.

Assistance to Women

Although the extreme persecution of women inflicted by the Taliban has ceased, women still have a difficult time in present-day Afghanistan.

- The U.S. has helped the Afghans to establish women's resource centers.
- Over one million dollars has been used for training women in business and for educating girls.

Transportation Infrastructure

The United States has committed $800 million to reconstruct the main transportation route between Kabul, Kandahar, and Herat. The project, underway since 2002, is scheduled for completion by 2004.

The United States is helping the Afghan government to repair and reconstruct bridges and 4,000 km of secondary and tertiary roads. The United States is also helping to winterize a major tunnel that is a key link to northern Afghanistan as well as participating in revitalizing over 6,000 water wells, canals, dams, and water systems in the country.

Child Welfare

- The United States has funded an immunization program for Afghan children. Since April 2002, 4.3 million Afghan children have been immunized against measles.
- America's Fund for Afghan Children has raised over $11.4 million to help Afghan children. Moneys from this fund have been used to purchase school supplies and books as well as teacher kits. Funds have been used to build playgrounds. Winter relief and health kits for children have also been supplied from this fund.

Continued on next page

Refugee Welfare

More than two million refugees and displaced persons have returned home to Afghanistan since the overthrow of the Taliban.

- The United States has donated funds and assistance to help resettle these refugees and to provide them with food and basic needs, as well as shelter kits to the neediest.
- An Afghan Conservation Corps has been established, giving Afghan refugees who have returned to the country and demobilized fighters employment and job training.

Health

- The central disease surveillance system for the eradication of polio was re-established, and a national curriculum for midwives was revised.
- Health facilities, including over 28 clinics and hospitals, have been rebuilt and refurbished. A national plan has been drawn up to determine the location and type of construction for an additional 600 health care facilities.

Economic and Agricultural Activities

The Afghan government has an ambitious business enterprise policy. The aim is to restore full central government authority over the nation's currency and fiscal system. The United States is assisting the Afghans in developing banking, trade, and fiscal and business reforms. More privatization of state owned banks and public sector organizations is also sought.

Seventy percent of the Afghan population is engaged in agriculture. The United States has funded assistance to Afghan farmers aimed at improving crop production. In the spring of 2002, the United States donated 7,000 tons of seed, fertilizer and technical support to over 40,000 farmers. During 2002, crop production increased 80% over 2001 production.

http://www.developmentgateway.org/node/134111/ is an excellent site on reconstruction in Afghanistan, showing what various coalition members are doing to help the country.

Islam: Basic Ideas

Introduction

You have probably heard the name ***Islam*** associated with Afghanistan. Most people in Afghanistan and in surrounding countries believe in Islam. The Taliban and Osama bin Laden believe in a radical form of Islam. In this section, you will learn about Islam, its main ideas, customs, and practices.

(http://thetruereligion.org/intro.htm - - a short introduction to Islam)

Definition of Islam

Islam means submission to the will of God in the Arabic language. The word *Islam* has the same root as the Arabic word *salaam*, which means peace. Muslims around the world often greet each other with the phrase "Salaam" in much the same manner as Jews say, "Shalom," as a greeting. Both salaam and shalom come from the same common language root, the three-letter root - - *s-l-m*.

(http://media.isnet.org/off/Islam/basics/index.html) – discussion of various "doctrinal" aspects of Islam by noted theologians.

Allah

Islam is not named after a person like Christianity, which was named after Jesus Christ, Buddhism after Buddha, and Confucianism after Confucius. The central focus of Islam is always God.

The basic tenet of Islam is that you must submit to Allah (the Arabic word for God) and live according to His divinely inspired law.

The Precept

- The key truth that Allah has revealed to mankind is that the only divine and worshipful "being" is Allah, the almighty God; thus all human beings should submit to and worship Allah.

- Allah has 99 names, among them, The Gracious, The Merciful, The Beneficent, The Creator, The All-Knowing, The All-Wise, The Lord of the Universe, The First, The Last, and others.

No matter what "sect" of Islam a person belongs to, he believes in this important religious belief that is the common thread of Islam.

Arabs and Islam

Muslims (Moslems)

Now that you know that the religion is called Islam and the name of the God of Islam is Allah, what does the word *Muslim* mean? The word has the same s-l-m root as the words *Islam* and *salaam*. A Muslim is a person who submits to the will of Allah, regardless of race, nationality, or ethnic background. A Muslim has to submit completely and obediently to Allah and live according to Allah's message.

- Muslims worship Allah alone and must not worship any person, place or thing other than Allah. They believe that Allah is the God for the Christians, the Jews, the Muslims, the Buddhists, the Hindus, and even atheists.

- Orally repeating the basic belief of a Muslim found in the "motto" of Islam: "*Laa Elaaha illallaah*" which means, "There is no god but Allah," is the way a Muslim professes that he belongs to Islam.

Importance of Arabic in Islam

The "motto" above, "*Laa Elaaha illallaah*," is in Arabic. Even though the people of Afghanistan speak a number of different non-Arabic languages, the language of Islam is Arabic, and this is the language used in "religious services." (Historically, Islam began among the Arabs of what is now Saudi Arabia.)

- The prayers, sayings, and many blessings of Islam are all spoken in Arabic, just as the holy book of Islam, the Qur'an (Koran) is written in Arabic.

- This common religious language of Islam is one of the ties that bind all Muslims together.

Continued on next page

Arabs and Non-Arab Muslims

Because of the importance of the Arabic language in Islam, many people assume that most Muslims are Arabs. This is not the case: About 80% of all Muslims are **not** Arabs.

- In fact there are more Muslims in Indonesia than in the Arab Middle East.
- Besides Indonesia, Muslims make up the majority in such non-Arabic countries as Turkey, Pakistan, Uzbekistan and other central Asian countries of the former Soviet Union.
- There is a large minority in China too.
- In Europe, Albania is Muslim. Bulgaria, Bosnia, and Macedonia have large Muslim populations. Elsewhere in Europe, there are immigrant communities of Muslims from Africa, Turkey, and Asia in France, Britain, and Germany.
- In the Americas Muslims have increased in recent years, both from conversions and immigration; 20% of the population of Suriname in South America is Muslim. Today there are about five million Muslims in America.

Muhammad (Mohammed)

Importance

Muhammad is … the Messenger of God
and the last of the prophets ... (Qur'an, 33:40)

This one phrase from the Qur'an shows Muhammad's place in Islam and in world religion. Muslims believe that Muhammad is the prophet of Islam. His message is Islam. The revelation of Islam that Muhammad received is the *Qur'an* (Koran).

Muhammad and Other Religions

In terms of other world religions, Muhammad is the last prophet of God to mankind.

- He is the last messenger of God.
- His message applies to all mankind, regardless of their beliefs.
- Muhammad is the successor to Moses, Jacob, Isaac, Abraham, and Jesus.

Muhammad is **the** prophet among all previous prophets and messengers.

- He reinterpreted and corrected the message of Allah and gave mankind the true word from Allah. In a sense, Muslims consider that he "cleansed" the message.

- In spite of his importance in Islam and for the world, Muslims do not worship him or ascribe any divine qualities to the prophet.

Muhammad, His Calling

Muhammad was born in Mecca (Makkah) in what is now Saudi Arabia, in the year 570 CE. His father died before he was born. Soon after his birth, his mother also died, so Muhammad was raised by his uncle.

Because of his trustworthiness and sincerity he was often asked to arbitrate disputes. He has been described as calm and meditative. Muhammad was very religious. It was his habit to meditate in a cave near Mecca.

At the age of 40, during one of his retreats, Muhammad received his first revelation from Allah through the Angel Gabriel. This revelation, which he continued to receive for twenty-three years, is known as the *Qur'an* (Koran).

Continued on next page

Spreading Islam

As Muhammad began to preach the truth that Allah had revealed to him, he and his small group of followers suffered persecution.

- This became so strong that in the year 622 CE Muhammad and his followers emigrated from Mecca.

- This event, called the *Hijra* ('migration'), when they traveled about 260 miles to the north to the city of Medina, marks the beginning of the Muslim calendar.

- After several years, Muhammad and his followers returned to Mecca, where they made peace with their enemies and established Islam.

- Before Muhammad died at the age of 63, most of Arabia was Muslim, and within a century of his death, Islam had spread to Spain in the West and as far east as China.

Five Pillars of Islam

Introduction

The five tenets are the foundation for all Islamic belief. They are the cornerstones of Islam. (http://www.usc.edu/dept/MSA/fundamentals/pillars/ – an introduction to the five pillars.)

Commitment

Called the Ash-Shahaadah, this is the "credo" of Islam:

"I bear witness that there is none worthy of worship except Allah and I bear witness that Muhammad is His servant and messenger."

All Muslims subscribe to and repeat this belief.

Prayer

Called As-Salaat in Arabic, this is the requirement that adult Muslims pray at five specific times during a twenty-four hour day to Allah.

- This is a ritual prayer requiring Muslims to say specific prayers and to coordinate hand and arm movements, eventually culminating in complete prostration or bowing down with the hands on the floor to show complete submission to Allah.

- While performing As-Salaat, Muslims face in the direction of Mecca.

Alms

Known as the Az-Zakaat in Arabic, this means that Muslims must contribute to the support of those less fortunate Muslims--support of the poor. Muslims are encouraged to pay their Zakaat during the holy month of Ramadan.

Continued on next page

Five Pillars of Islam, Continued

Fasting

In Arabic this is called As-Sawam. It means that during the month of Ramadan, adult Muslims who are in good health

- Cannot eat, drink, or smoke from sunrise until sunset every day of this holy month.

- Must abstain from sexual relations.

The intention of fasting is to worship Allah. Ramadan is a sacred month because it commemorates the time when Allah first revealed the Qur'an to Muhammad. The time for Ramadan changes each year because it is based on a lunar calendar. Ramadan comes 11 days earlier each year.

Pilgrimage

Called Al- Hajj (Hadsh), this pilgrimage requires a Muslim in good health to travel to Mecca in Saudi Arabia at least once in a lifetime.

- Specific preparation is necessary for the Hajj; for example there are rules about shaving and cutting hair and what to wear.

- One of the high points of the Hajj is the *umrah*, which means performing a ritual walk seven times around the sacred stone of Islam called the Kab'ah.

- The Hajj commemorates the sacrifices and faith of Abraham, his second wife, Hagar, and their son, Ishmael.

According to the Council on Islamic Education, it is the largest, regularly scheduled international gathering on Earth.

Al Qur'an (Koran) and Sunnah

Definition

Al Qur'an (the Qur'an) is the revealed word of Islam, the holy book of the Muslims.

- Muhammad received the message in bits and pieces from the Angel Gabriel during a period of 23 years.
- The word *Al Qur'an* means the recitation. During Muhammad's lifetime it was recited publicly.
- The original language of the Qur'an is Arabic.

It is important to remember that Muslims believe the Qur'an is **the** word of Allah, untouched and uninterpreted by human beings. As Allah's final message to mankind, it is the holy book that supercedes all others: the *Old Testament*, the *Gospels*, etc.

Form and Content

The Qur'an is a collection of surahs or chapters, many of which you might call verses.

- Muhammad received the surahs from Allah. Scribes selected by Muhammad usually wrote down these verses. Sometimes they wrote on wood, trees, parchment, and even stones. Many followers also memorized the Qur'an by heart.

- In later years, the Qur'an was recopied and refined; accents and markings for reading were added.

Interpretation

- The Sunnah is the spoken word and acts of Muhammad. The Sunnah, Muhammad's actual words and deeds, uses the life of Muhammad to explain and expand on the verses and teachings of the Qur'an so that Islam can eventually become a world religion.

- A second form of interpretation is called hadith. This is narration about the life of the prophet or what he approved - as opposed to his life itself, which is the Sunnah.

The Qur'an, the Sunnah, and the ahadith (plural of hadith) are the basic body of Islamic religious doctrine for all Muslims. (http://www.afghan-network.net/Islam/) (http://www.usc.edu/dept/MSA/)

Jihad (Djihad)

Definition

The word *jihad* means struggle in Arabic. This means a struggle between the forces of Allah (good) and the forces of evil --an eternal struggle.

You have probably heard the word *jihad* in reference to Osama bin Laden and the Taliban. They consider the terrorism that they perpetuate a jihad. Actually, the word *jihad* has a couple of interpretations.

Self-defense

Islam allows fighting in self-defense and in defense of religion. There are rules of combat including prohibitions against harming civilians and against destroying crops, trees and livestock.

If good Muslims do not risk their lives in the cause of Islam, injustice would triumph. The Qur'an says:

> " *Fight in the cause of God against those who fight you, but do not transgress limits. God does not love transgressors.*" (2.190)

War is the last resort and is subject to the precepts of Islamic sacred law.

Inner Struggle

The second and more personal meaning of jihad is an inner struggle. Each of us, according to Islam, is constantly fighting a "war" against self-centered desire, and egotism, with the final goal of attaining inner peace.

Islamic Schools (Sects)

Religious vice Secular Law: Western View

The United States and most other Western countries make a clear distinction between the rule of law and religion.

- Religion is a separate personal issue that does **not** determine law.
- A legal case may make use of religion, but religion is not the determining factor in most legal cases.
- Simply put, a religious denomination, such as Methodism, Catholicism, or Judaism, doesn't have the power to rule on a case of adultery in a civil court or on a case of fraud in a criminal court.

The United States has a secular system of law; religion plays no direct part in such a system.

Islamic View

If you look at Islam, the perspective is different. Islamic law consists of guidelines and rules that determine all aspects of a Muslim's life from how to perform ritual prayer to conducting business transactions.

- It also includes crimes and the appropriate punishments for each.
- These Islamic laws are based primarily on the Qur'an and are called the Sharia.
- Muslims must apply these precepts handed down to Muhammad in the sixth century to present-day situations.

Interpretation and Re-interpretation

This kind of religious legal system, particularly how to interpret the Qur'an and the ahadith, has resulted in a certain degree of divisiveness in Islam, just as varied interpretations of Christian doctrine have caused Christianity to split into various denominations. The two main sects of Islam, which are the schools of Islam, are the

- Sunni
- Shi'ites (Shi'ia)

Continued on next page

Shi'ites (Shi'ia)

The basic divisiveness in Islam is based on an historical incident:

- After Muhammad died, there was confusion about who should succeed him as the leader of Islam. One group felt that the successor was Imam Ali who had been appointed by Muhammad according to Allah's decree. Ali was a relative of Muhammad and the first to accept Islam.
- This group, mostly from the household of the prophet, was a minority and became known as the Shi'ites. They also believe that the leader of Islam must be endowed with grace and benevolence and should be infallible.

Sunnis

The majority group, called Sunni Muslims, believe that:

- Muhammad did not choose a specific successor and probably assumed that after his death, Muslims would find their own leader.
- The prophet did not tell his followers how they ought to select their leaders or what qualifications their leaders should have.

The Sunnis chose a leader, later called the caliph, from outside Muhammad's household. The Shi'ites refused to accept the Sunnis' choice and split off from this main group.

Results of the Division

Because of this succession split, variations in Islamic doctrine, law, and practice have developed between these two main groups over the centuries.

- Sunni Muslims predominate in Saudi Arabia, Afghanistan, Pakistan, Turkey, and Indonesia.
- Shi'ites are the majority in Iran and southern Iraq.

Sunnis make up about 83 percent of Muslims, according to the *Encyclopedia Britannica*; Shi'ites, about 16 percent; and a few other small groups, the remainder.

Osama (Usama) bin Laden and Islam

Bin Laden's Religious Background

Bin Laden's brand of Islam is based on extremely conservative Islam called Wahhabism, which was espoused by Muhammad bin Abd al-Wahhab (1703-87). He was the founder of the sect and the co-founder of Saudi Arabia. The Wahhabis and the royal Saud family have long intermarried. The Saudis have adopted many of the conservative precepts of Wahhabism.

Wahhabism: The Way of a Muslim Life

Wahhabi Islam is a total religious system. It has an answer for every question about a Muslim's life including social, legal and spiritual aspects. Wahhabi Islam is ascetic.

- Men should wear short robes and even avoid the black cords used on turbans and headgear.
- Mosques should have no decoration.
- Drinking alcohol is forbidden.

Punishment is based exclusively on the Qur'an, for example:

- The right hand should be amputated for theft.
- Adulterers should be stoned to death.
- Murder and sexual deviation are punishable by beheading. To this day Saudi Arabia uses these punishments, especially beheading for capital crimes.

Bin Laden and the Taliban

The kind of radically, conservative state that the Taliban had developed in Afghanistan had roots in Wahhabism, which inspires and feeds on the Taliban's fundamentalist religiosity.

- For Islamic fundamentalists, the Taliban have created what approaches a pure, ideal society based on Islamic religion and law.
- This society fits bin Laden's mold too.

The Taliban allowed bin Laden to remain in Afghanistan as a guest. They denied consistently that he had anything to do with the terrorist acts of 9/11 and refused to extradite him as requested by the US government. They continuously claimed that they did not know of his whereabouts.

Continued on next page

Focus on Expansion and Terrorism

 While bin Laden follows the conservative Wahhabi tenets, what he most fervently supports is that his brand of Islamic faith must expand. (To support this kind of expansion, Saudi Arabia supported the Afghan *mujahidin* when they were fighting the Communists. So did Osama bin Laden.)
http://www.newyorker.com/FROM_THE_ARCHIVE/ARCHIVES/?010924fr_archive03 – good article on bin Laden, his life and religion

 Now bin Laden regards the struggle as a war between two civilizations: His Islam and the non-Muslim civilization, specifically the United States. Toward winning this struggle, bin Laden has committed his forces to terrorism.

http://www.brookings.org/fp/projects/terrorism/faqs.htm#qb4 – a set of informative questions about terrorism.

Bin Laden-sponsored Terrorism

 Besides the infamous 9/11 terrorist attacks against the Pentagon and the twin towers, bin Laden allegedly sponsored the attacks on American embassies in Africa.

 U.S. and Russian officials are also concerned that he is financing Chechen rebel operations out of Dagestan, a former Soviet republic in Asia.

A primary concern is that bin Laden's forces are attempting to acquire or build chemical and/or biological weapons to use in their terrorist attacks. In the case of the Chechen rebels, he may be trying to help them acquire radiological dispersal devices.

Continued on next page

Al Qaida (Al Qaeda)

Al Qaida (Al Qaeda) is the name of the network of Islamic extremists that bin Laden has at his command to carry out his radical Islamic terrorism.

- Al Qaida consists of a group of about 3,000 commanders. Troops of Afghanis and Pakistanis number 200,000+.
- There are a number of terrorist centers, with numerous cells, in North America, Yemen, Saudi Arabia, Albania, Kosovo, Algeria, Chechenya, all the former Soviet republics of Central Asia, the Philippines, Egypt, Ethiopia and Somalia.

Bin Laden's "Army"

Al Qaida is bin Laden's personal "army." Bin Laden could probably rally an army of over 100,000+ men around the world, excluding those currently "in service" in Pakistan and Afghanistan.

- Many members of his overseas forces are ready for both sustained and one-time operations.
- In the interim, the members of these terrorist forces return to normal "civilian" life in host countries. They are also available for terrorist operations on their own countries.

Funding for Al Qaida (Al Qaeda) Terrorism

The primary sources for funding Al Qaida-sponsored terrorism have been:

- The sale of drugs, primarily opium.
- Contributions from Saudi Arabia through various banks and from various Saudi businessmen.
- What might be called tribute money to Al Qaida from countries that do not want the organization to establish cells in their own countries.
- Funds from bin Laden's personal fortune.

Continued on next page

Bin Laden and Saudi Arabia

Even though the country where he grew up, Saudi Arabia, is extremely conservative, bin Laden finds that it is not conservative enough. He thinks the following reasons, which are contrary to his purist concept of Islam, are grounds for overthrowing the Saudi royal family:

- Alliance of the ruling Saudi family with the West
- Saudi Arabia's dependence on American and other foreign troops who came to the country to defend it during the Gulf War
- Corruption of the Saudi regime

Appendix A: Short History of Terrorist Activity

Timeline The Activities of the Al Qaida

The Al Qaida has a long history of terrorist activity, which dates back to the early 1990s. Many of their earlier operations failed to achieve the desired effect; however, experience has enabled the organization to develop a more sophisticated approach.

- Suspected in the bombing of night club in Bali
- Suspected in September 11 plot to hijack planes and destroy the World Trade Center, Pentagon, and other unspecified targets (2001)
- Plotted an unsuccessful effort to attack U.S. and Israel tourists during millennial celebrations (2000)
- Implicated in bombings of U.S. embassies in Nairobi, Kenya and Dar es Salaam, Tanzania that killed over 300 (1998)
- Accused by the U.S. of bomb attacks on American military personnel in Dhahran, Saudi Arabia (1996)
- Linked to unsuccessful plot to assassinate President Clinton in the Philippines (1995)
- Linked to unsuccessful plot to bomb multiple U.S. trans-Pacific flights (1995)
- Linked to unsuccessful plot to bomb U.S. and Israeli embassies in Manila (1994)
- Linked to unsuccessful plot to assassinate Pope John Paul II in Manila (1994)
- Claims to have downed U.S. helicopters and killed U.S. servicemen in Somalia (1993)
- Attempted bombing of U.S. troops in Aden, Yemen (1992)

Source: U.S. State Department

Appendix B: International Associations

Top Commanders

Al Qaeda is an international association of allied groups operating in many parts of the world. Its top commanders are

- AYMAN AL ZAWAHRI, head of the Egyptian Jihad Islami, who is Bin Laden's senior deputy and heir apparent
- JUMMA MAMANGANI, an Uzbek, who was recently appointed Al Qaeda chief of operations. He is former commander of the Moslem Army for the Liberation of Kyrgystan. Three key Afghan training camps, at Jalalabad, Farmada and Daronta, are under his command.
- FATEH KAMEL, who leads the most militant cells of the extremist Algerian GIA. In the name of Al Qaida, he controls terrorist cells in the United States, Canada and Algeria.
- MUHAMMED ATIF aka SUBHI ABU-SITTAH, who is nominal chief of staff of the network and its brightest military brain. He comes from the Egyptian Jihad Islami
- IMAD MUGHNIYEH, the former Lebanese Hizballah hostage-taker and bomber, who is in charge of the combined terrorist campaign around the Middle East, the Persian Gulf, Europe and Israel.

Throughout the World

On top of the 3,500 hard core commanders and 110,000 Afghan and Pakistani troops, Al Qaeda retains another 6400 commanders in 12 centers: North America, Yemen, Saudi Arabia, Albania, Kosovo, Algeria, Chechenya, Tadjikistan and all the former Soviet republics of Central Asia, the Philippines, Egypt, Ethiopia and Somalia.

The numbers Bin Laden can muster differs from place to place. In North America, together with his closest ally, the Egyptian Jihad Islami, some 2,500 hard core fighting men; in Yemen, where his family originated before migrating to Saudi Arabia – 2000 directly. But Bin Laden has a special relationship with the commanders of the 20,000-man strong irregular "Moslem Liberation Army, which hold a monopoly of the arms trade of the Arabian Peninsula, the Horn of Africa, the Red Sea and East Africa. Its reach sometimes goes as far as Iran.

Bin Laden often serves as the MLA's clearing bank and ready bankroll for arms deal.

Continued on next page

Saudi Arabia

In Saudi Arabia, where the privileged and affluent bin Laden clan lives, Osama commands some 200-250 hard-core commanders, but there are many more potential partisans among the disaffected tribes in the central and eastern provinces, especially the Nejd, as well as in the Saudi armed forces and national guard.

An intensive Al Qaeda recruitment effort in those Saudi forces could cause their collapse and drop in Bin Laden's lap their arsenals, with some of the most sophisticated hardware in use today.

Overseas Legions

According to conservative estimates, the millionaire-terrorist could most probably rally around the world roughly the same number of fighting men as those flocking to his flag in Afghanistan and Pakistan, namely an army of over 200,000 men. Many members of his overseas legions are available for both sustained and for one-time operations. In between, they simply go back to their normal pursuits and their homes in host countries. They are also available for terrorist operations on their home ground.

Appendix C: Other References

Resources

Besides the websites and URLs cited in the body of this handbook, the following resources will help you better understand Afghan culture and people:

- *Encyclopedia of World Cultures* (particularly volume III), edited by Paul Hockings. Boston: G.K. Hall and Co., 1992. Excellent summaries of the different tribal and ethnic groups of Afghanistan and neighboring countries.
- *The Europa World Year Book 2001* (volume I). London: Europa Publications, 2001. An excellent in-depth survey of Afghanistan.
- Friedman, Norman. *Terrorism, Afghanistan, and America's NEW WAY OF WAR.* Annapolis: Naval Institute Press, 2003.
- Hoffman, Frank G. (LtCol, USMCR, Ret.). "The U. S. Marine Corps in Review." *Proceedings*, May 2002: Vol. 128/5/, 84-90.
- Hoffman, Frank G. (LtCol, USMCR, Ret.). "The U. S. Marine Corps in Review." *Proceedings*, May 2003: Vol. 129/5/, 96-102.
- Marsden, Peter. *The Taliban: War, religion and the new order in Afghanistan.* New York: Zed Books Ltd., 1999. In-depth study of the Taliban including their origins and type of leadership, their religious "creed," implementation of religious beliefs, etc.
- O'Balance, Edgar. *Afghan Wars 1839-1992: What Britain Gave Up and the Soviet Union Lost.* New York, Brassey's, 1993. Good account of the major conflicts in Afghanistan over the years: the Anglo Wars, the Soviet period, and civil wars.
- Rubin, Barnett R. *The Search for Peace in Afghanistan: From Buffer State to Failed State.* New Haven: Yale University Press, 1995.

Made in the USA
Monee, IL
04 November 2026

67377579R00039